Fun math book

MY FIRST NUMBERS

from 0 to 10

Author: Irina Korolenko

FunMathtic!

AF887698

Fun Math Book is filled with creative math tasks, logic and visual puzzles, problem-solving tasks and even artistic exercises. Not all of the tasks in the book have a definite answer. Sometimes your child needs to find his or her own reasoning and way to solve it.
We see our tasks and exercises as an invitation to think and to create.

If you like to find more math materials like this, visit our on-line store:
https://www.teacherspayteachers.com/Store/Funmathtic

Contents

Lesson 1: One Going to the beach Date:

1 One

1 1 1 1 1 1 1 1 1 1 1 1 1 1 1

One One One One One One One

Circle the objects that represent one thing:

Continue the pattern:

Fill the scene:

Add one child, one kite, one shell and the sun.

What else belongs on the beach?

2 Two

2

Two

What pairs of things you can find on the boy below?

My name is

Draw a picture of yourself

Can you figure out how to continue the pattern?

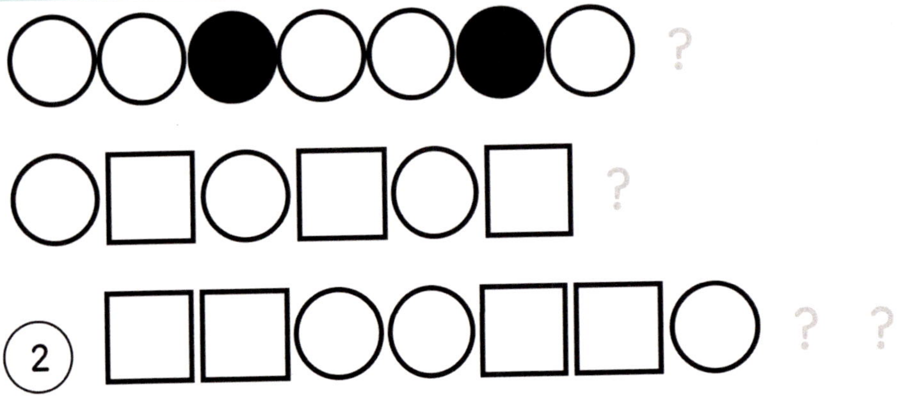

Trace the numbers:

1 1 1 1 1 1 1 1 1

1 1 1 1 1 1

1 1 1 1 1 1 1 1 1 1 1

2 2 2 2 2 2 2 2 2

2 2 2 2 2 2

2 2 2 2 2 2 2 2 2 2 2

Which kite is whose?

Trace and write the numbers:

1 2 1 2 1 2 1 2 1

2 1 2 1 2 1

1 1 2 2 1 1 2 2 1 1 2

2

1

2

1

3 Three

3 3 3 3 3 3 3 3 3 3 3 3 3 3

Three Three Three Three Three

Circle the planets which have number 3 on them.

Finish the picture – add the details listed below:

Add to the starship
and to the space around:

 – 3

 – 3

 – 2

What else can you add to the picture?

Take a break – color the picture.

Lesson 4: Four

Animals

Date:

4 Four

4 4 4 4 4 4 4 4 4 4 4 4 4

Four Four Four Four Four Four

Find all the animals that have four legs:

Find the odd one out: In each set, cross out the picture that is different from the others.

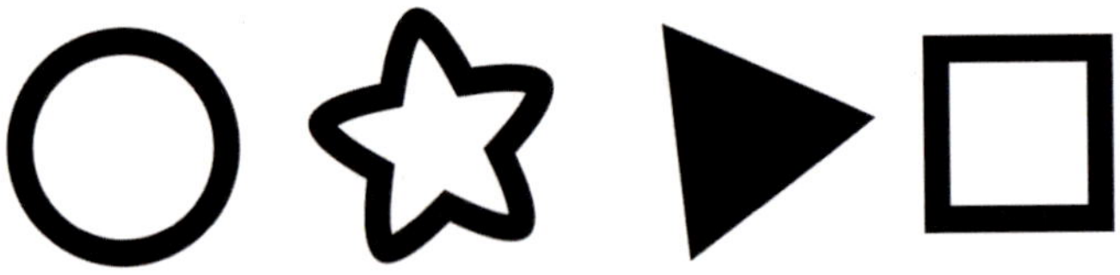

Trace the numbers:

3 3 3 3 3 3 3 3 3

3 3 3 3 3 3

3 3 3 3 3 3 3 3 3 3 3

4 4 4 4 4 4 4 4 4

4 4 4 4 4 4

4 4 4 4 4 4 4 4 4 4 4

Help the cat get to the fish.

Help the dog get to the bone.

Trace and write the numbers:

3 4 3 4 3 4 3 4 3

4 3 4 3 4 3

4 4 3 3 4 4 3 3 4 4 3

4

3

4

3

Date:

5 Five

5 5 5 5 5 5 5 5 5 5 5 5 5 5

Five Five Five Five Five Five

Color all the jars that contain five pieces of fruit or vegetables:

What will go next? Continue the pattern:

Draw 5 pieces of your favorite food

Lesson 6: Six

The City

Date:

6 Six

6 6 6 6 6 6 6 6 6 6 6 6 6 6

Six Six Six Six Six Six Six

Find all the buildings with six windows:

What can you easily find in a city and what is not so easy to find?

Color the picture by numbers.

1 – brown 2 – gray 3 – yellow 4 – blue 5 – red 6 – green

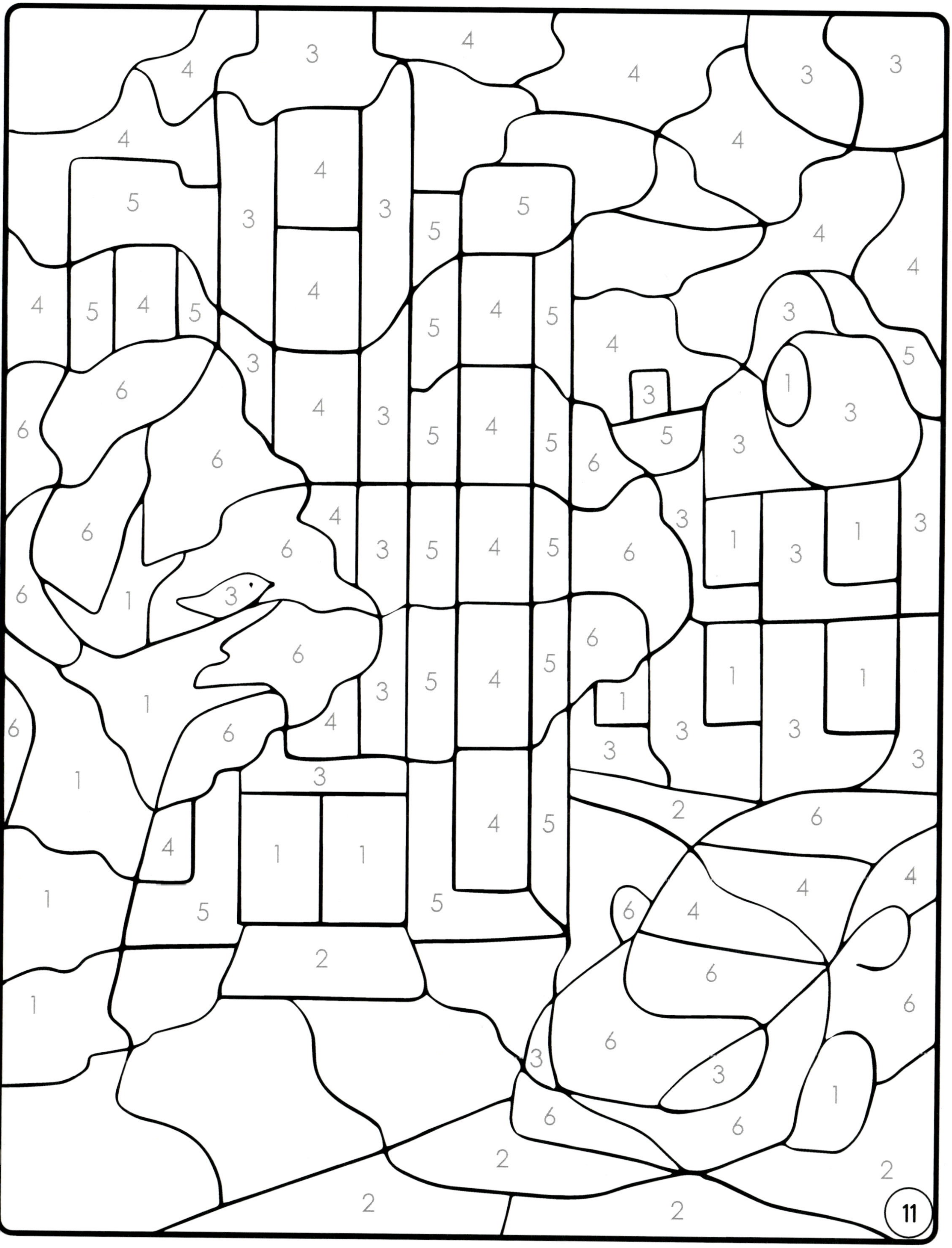

7 Seven

7 7 7 7 7 7 7 7 7 7 7 7 7

Seven Seven Seven Seven Seven

Find and circle all the trays with seven items on them:

Find two cups that are exactly the same:

Tea party! Connect the cups with the matching number of cakes:

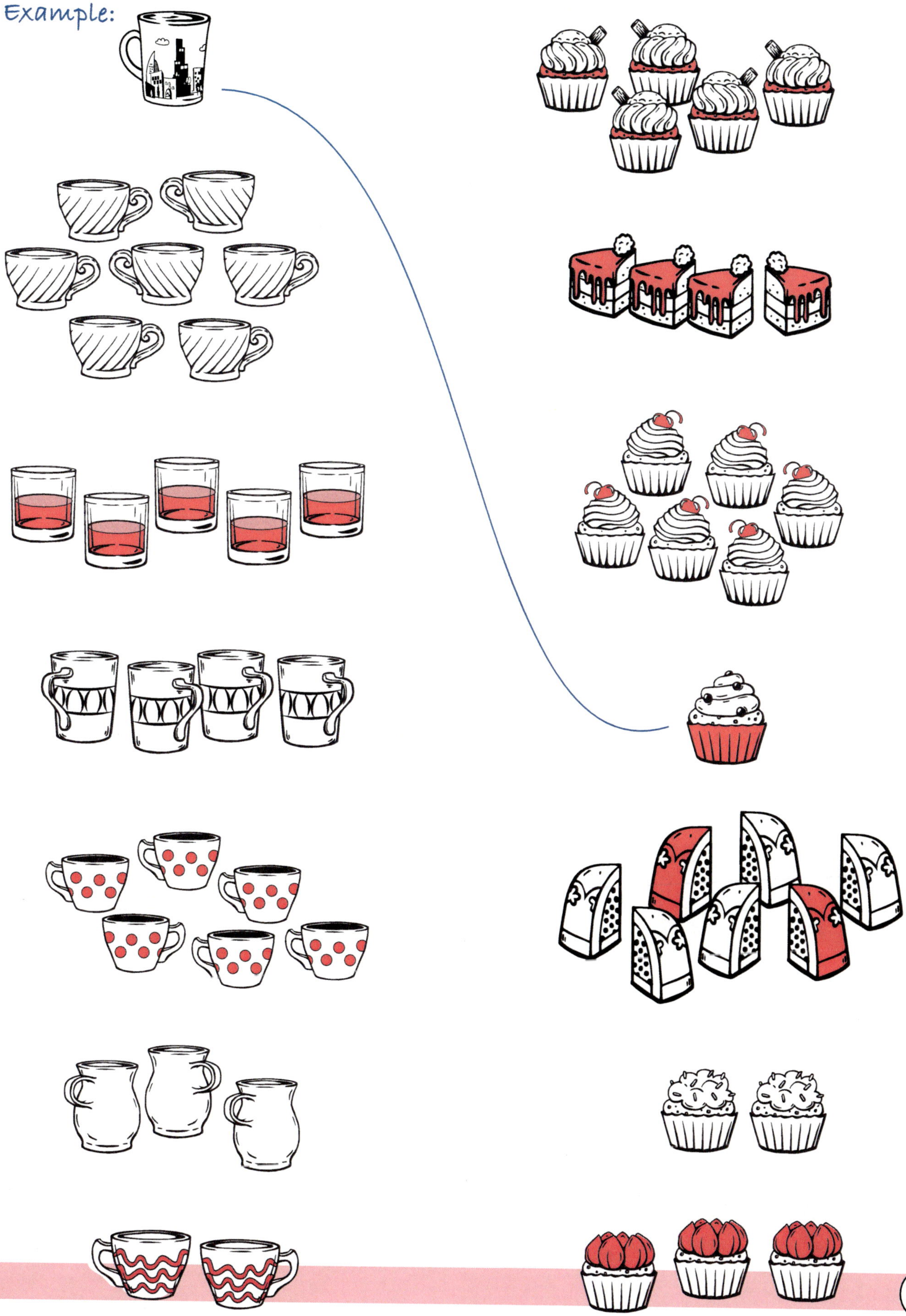

Find the missing piece:

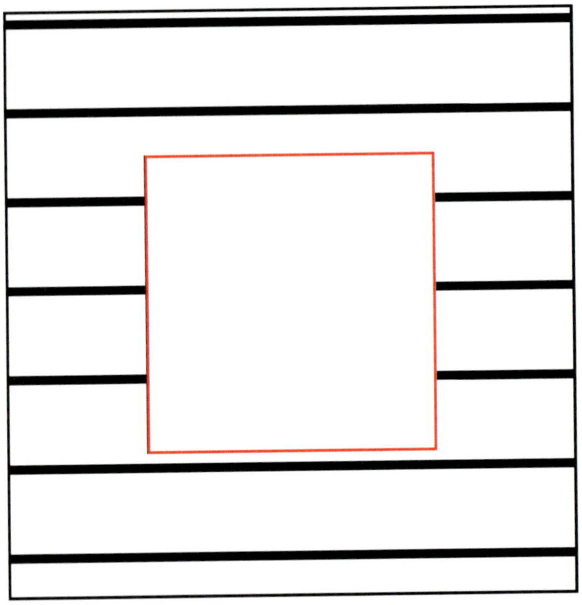

A)

B)
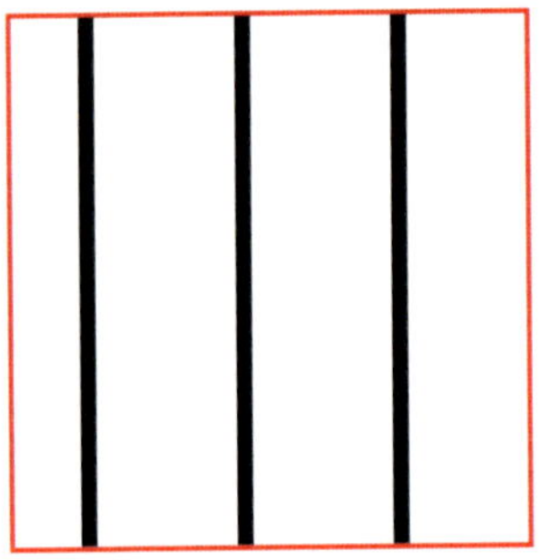

C)

Trace the numbers:

5 5 5 5 5 5 5 5 5   6 6 6 6 6 6 6 6 6

6 6 6 6 6 6   5 5 5 5 5 5

5 5 5 5 5 5 5 5 5 5 5   6 6 6 6 6 6 6 6 6 6 6

Find the missing piece:

A)

C)

B)

D)

Trace the numbers:

7 7 7 7 7 7 7 7 7 7 7 7 7 7 7 7 7 7

7 7 7 7 7 7 7 7 7 7 7

7 7 7 7 7 7 7 7 7 7 7 7 7 7 7 7 7 7 7 7 7 7

Find the missing piece:

1) 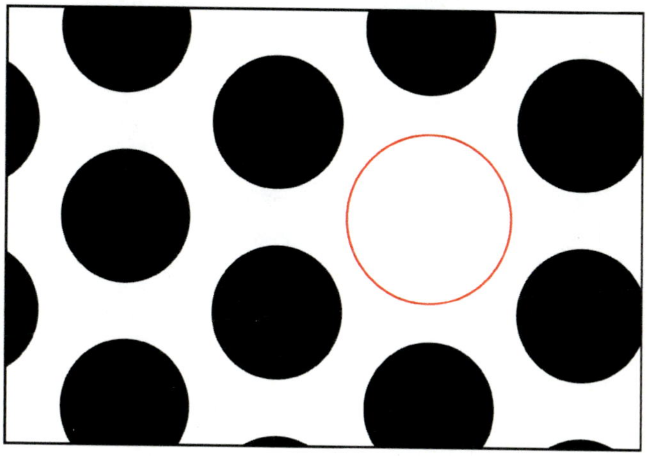

A) 

B) 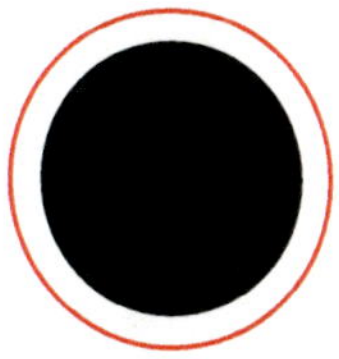

C) 

2) 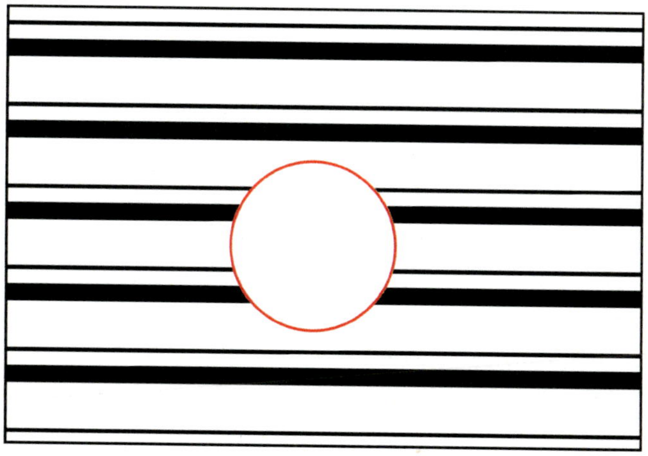

A) 

B) 

C) 

Trace the numbers, then continue the pattern:

5 6 7 5 6 7 5 6 7

7 6 5 7 6 5

5 5 6 6 7 7

8 Eight

8 8 8 8 8 8 8 8 8 8 8 8 8 8

Eight Eight Eight Eight Eight

Find all the trains that have eight wagons in them:

Can you find a cat and a strawberry?

Find the odd one out:

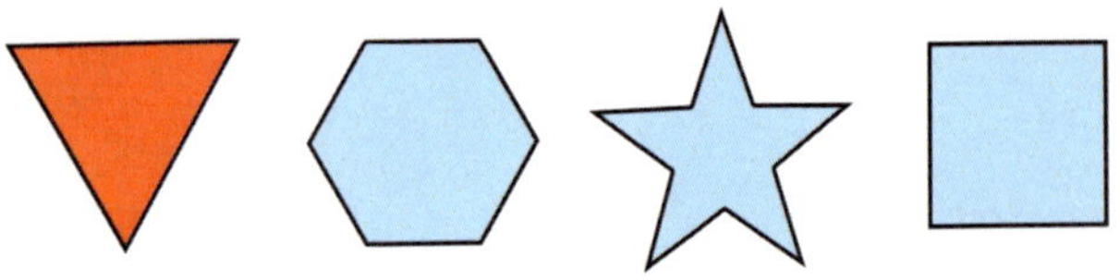

A K 5 E

Color the picture by numbers

4 – 5 – 6 – 7 – 8 –

Lesson 9: Nine   Christmas

Date:

9 Nine

9 9 9 9 9 9 9 9 9 9 9 9

Nine Nine Nine Nine Nine Nine

Color nine lightbulbs on every string:

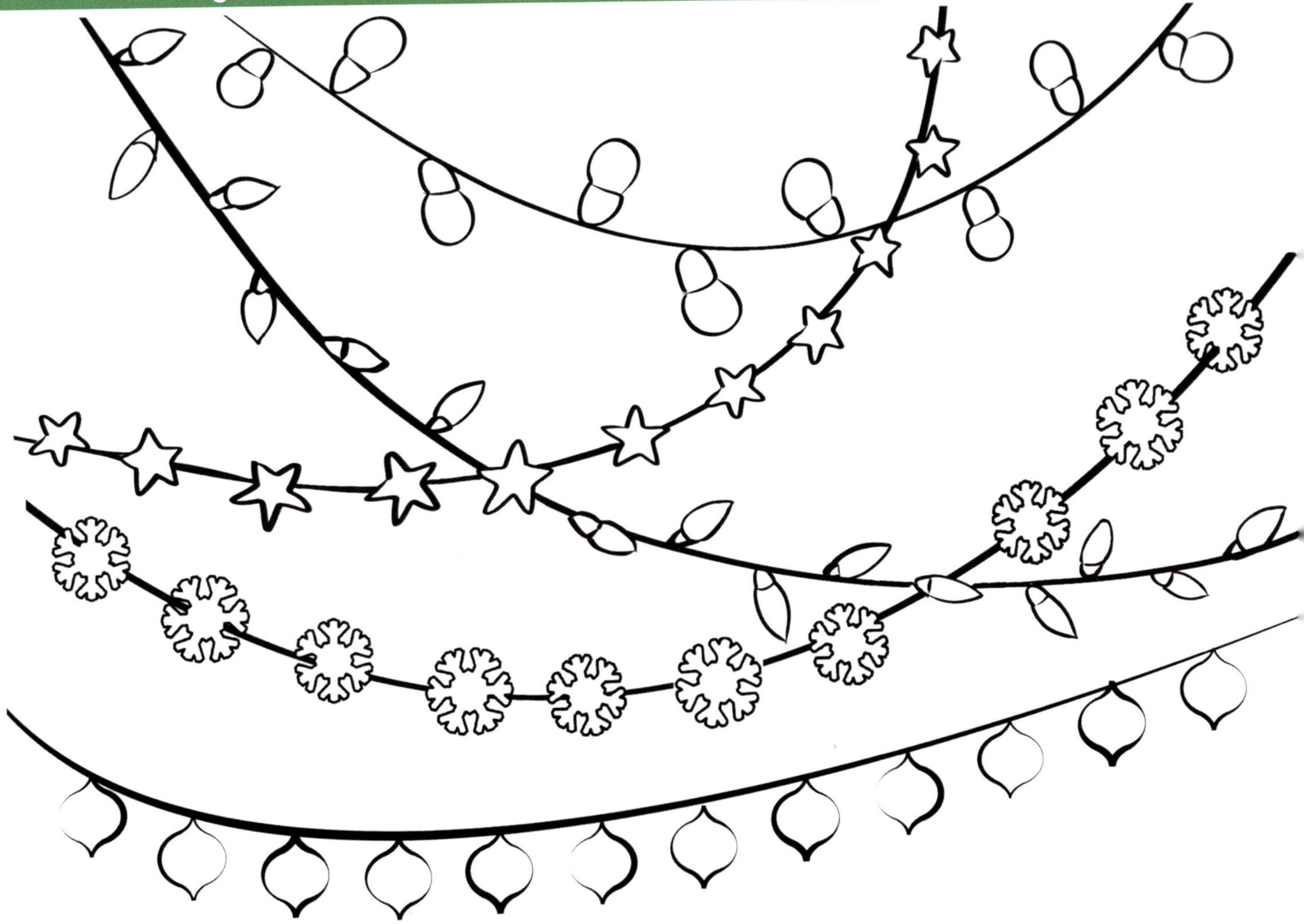

Continue the pattern:

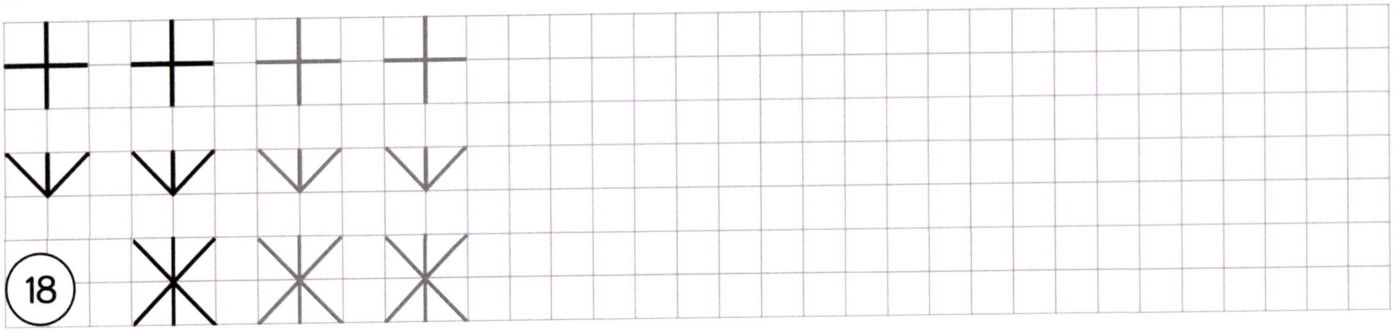

There are 9 kitties hidden on this Christmas coloring page. Can you find them all?.

Trace the numbers:

8 8 8 8 8 8 8 8 8 8 8 8 8 8 8

9 9 9 9 9 9 9 9 9

8 8 8 8 8 8 8 8 8 8 8 8 8 8 8 8 8 8 8

Count the numbers.

Little Mila dropped a box of numbers and all the numbers got messed up.
Can you help Mila to count the numbers?
She found three 'ones.' How many other numbers are in the pile?

1 - 3
2 - ?
3 - ?
4 - ?
5 - ?
6 - ?
7 - ?
8 - ?
9 - ?

Trace the numbers and finish the line:

9 9 9 9 9 9 9 9 9 9 9

8 8 8 8 8 8 8

9 9 9 9 9 9 9 9 9 9 9 9 9

Find the missing piece for each picture:

10 Ten

10 10 10 10 10 10 10 10 10

Ten Ten Ten Ten Ten Ten Ten

Add the missing numbers to the bead strings:

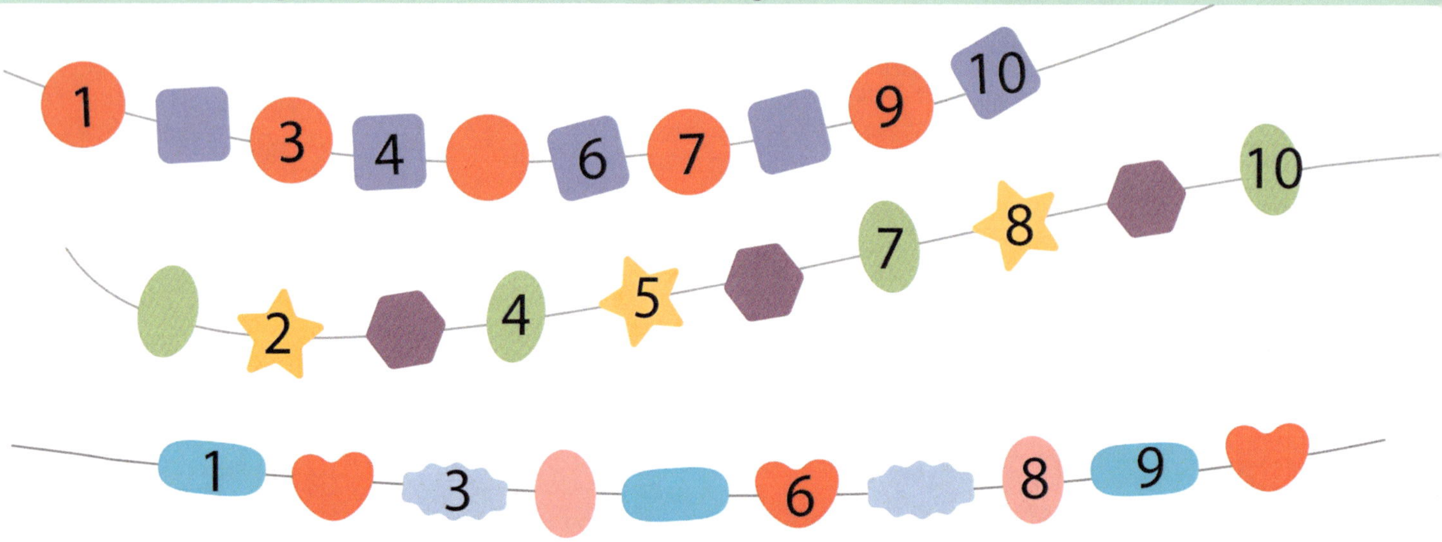

How many? Count the objects and write the answer.

Example  = 5

 = ?

 = ?

 = ?

 = ?

 = ?

Alex has beads with numbers on them:

He made three beautiful necklaces.
His younger sister dropped all of them and most of the beads fell off.
Luckily, Alex wrote down the numbers for every necklace.

Can you help Alex make the necklaces again? Would you like to color them?

Necklace 1: 1 - 7 - 1 - 7 - 1 - 7 - 1 - 7

Necklace 2 : 1 - 0 - 5 - 1 - 0 - 5 - 1 - 0 - 5

Necklace 3 : 3 - 6 - 3 - 9 - 3 - 6 - 3 - 9 - 3 -

Can you make your own necklace from these beads? Can you write the number pattern for it?

My necklace:

0 Zero

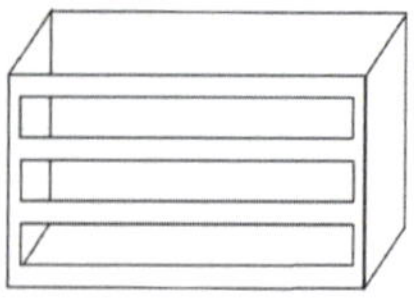

0 0 0 0 0 0 0 0 0 0 0 0 0 0

Zero Zero Zero Zero Zero Zero

Find and circle all the storage containers with zero items in them (the empty ones).

Continue the pattern

There are lots of zeros on this page!
Can you turn them into some nice pictures?

There are hidden numbers on this forest coloring page!

Can you find all the numbers from 0 to 9?

Lesson 12: Numbers 0 to 10 Bees

Date:

Connect each group of bees with the flower that has the matching number:

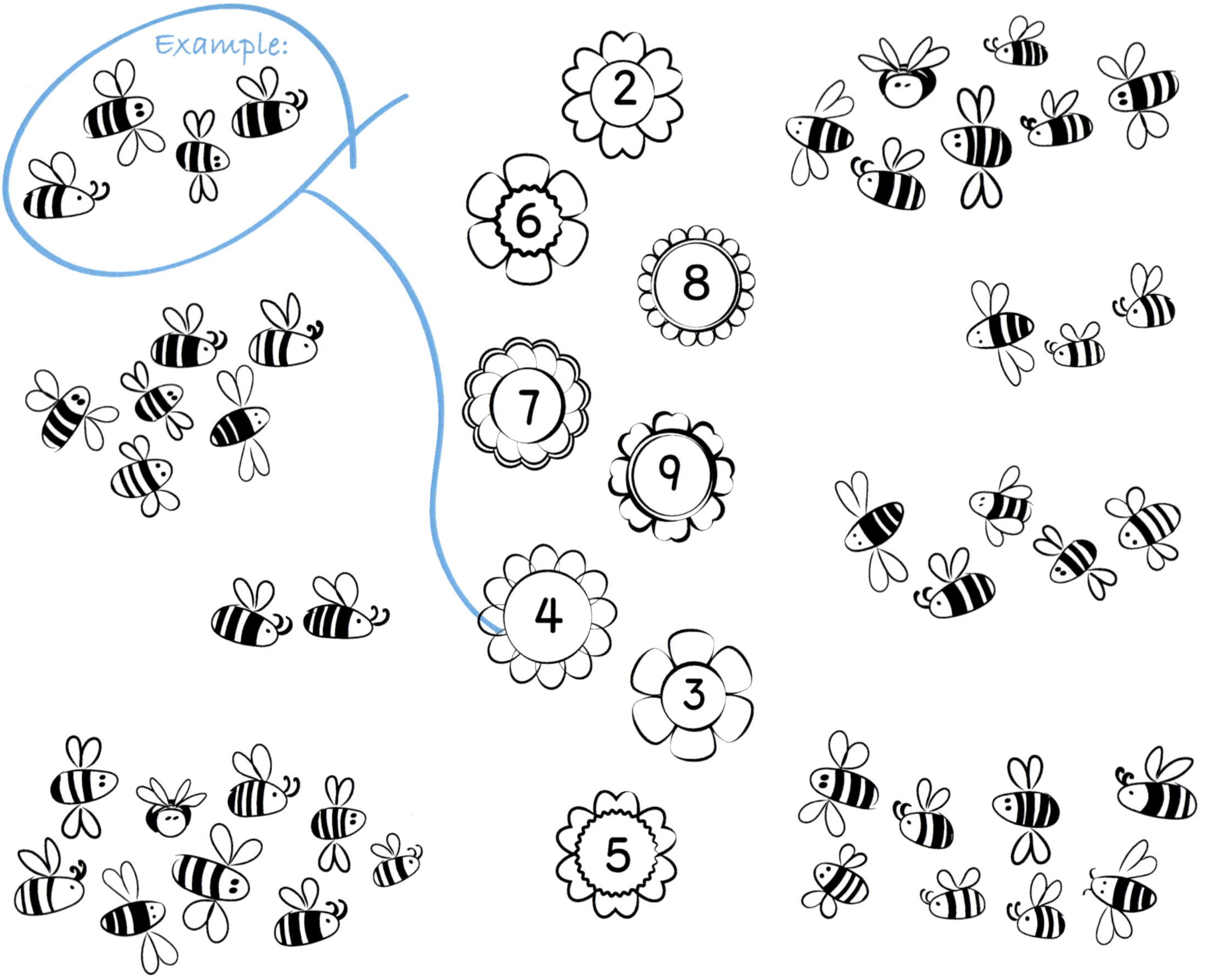

Find two jars of honey that are exactly the same:

Fill every honeycomb with numbers from 0 to 10.

Numbers can't repeat in one cluster

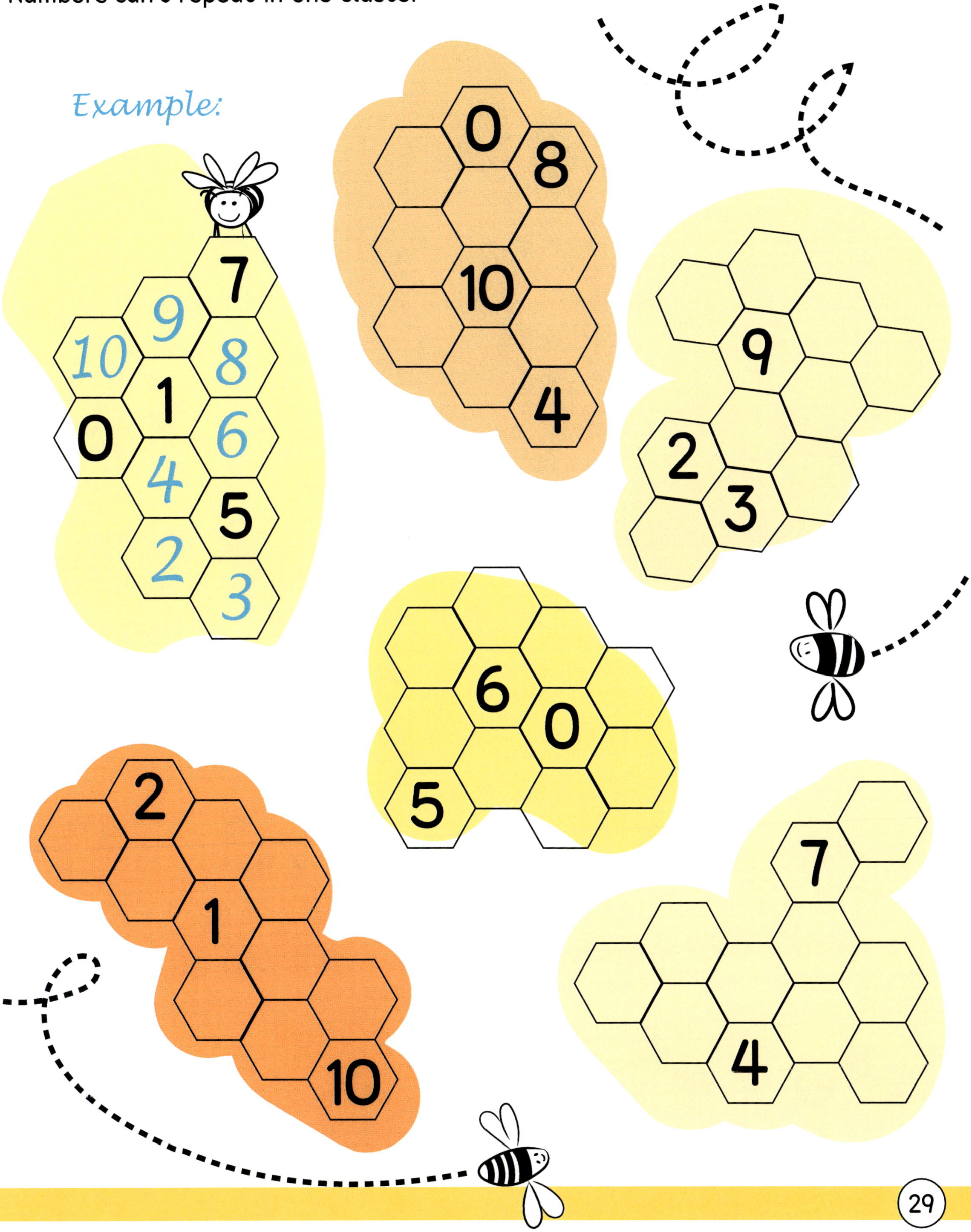

Lesson 13: More or Less

Date:

The hungry crocodile always wants to eat a bigger pile:

So he opens his mouth towards the plate with MORE vegetables on it:

We don't need to draw the crocodile to remember how to mark bigger number over the smaller one. We use signs:

More $>$

Less $<$

Which has more? Put more (>) or less (<) signs between the groups of fruit:

Lesson 14: Equal Aliens 

Date:

If we have the same number of things on both sides, we use the equal sign "="

 = 3 and  = 5

Fill the lines with these signs: > , < and =

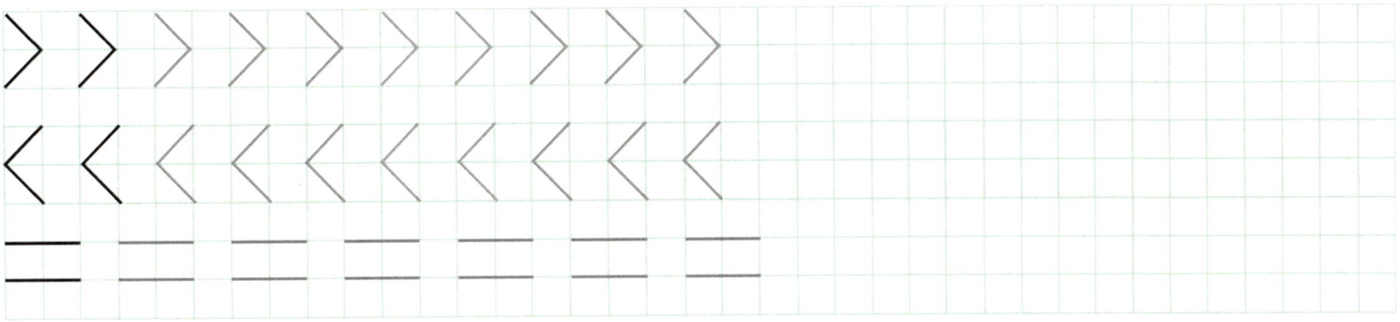

All these aliens need glasses. But which alien should get which glasses?

Count the eyes and find the eyeglasses with the equal number of lenses in them.

More, less or equal?

Between the groups of the fruits and veggies, put one of these signs: >, < or =

 >  Example

 ? 

 ? 

 ? 

 ? 

 ? 

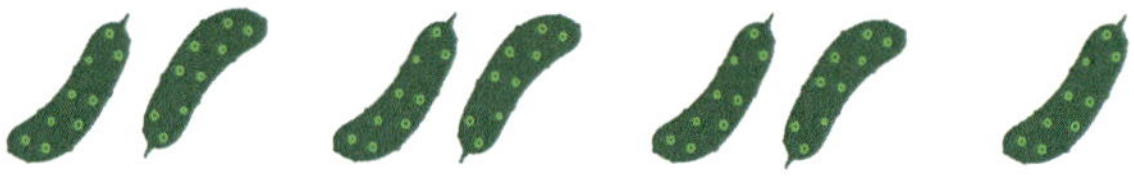 ? 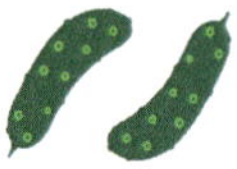

 ? 

Can you do it with numbers? Put the sign >, < or =

 ? 6

 ? 7

 ? 8

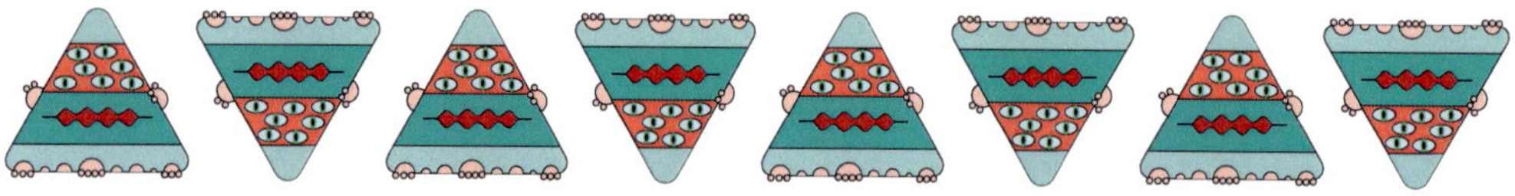 ? 9

 ? 10

Color the picture. How many balloons can you find here?

Lesson 15: Numbers 0 – 10   Castles

Date:

Fill in the missing numbers:

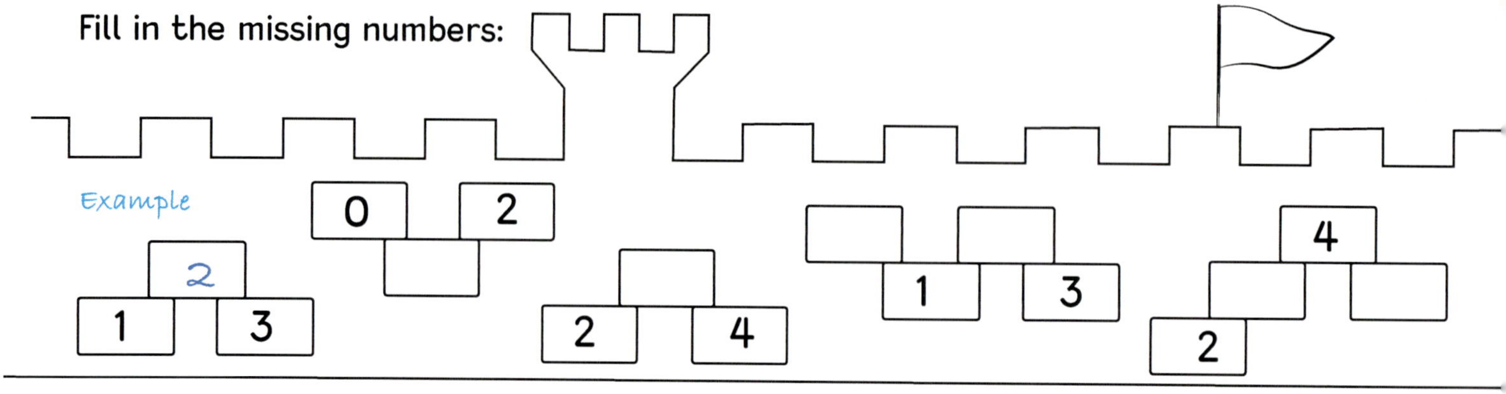

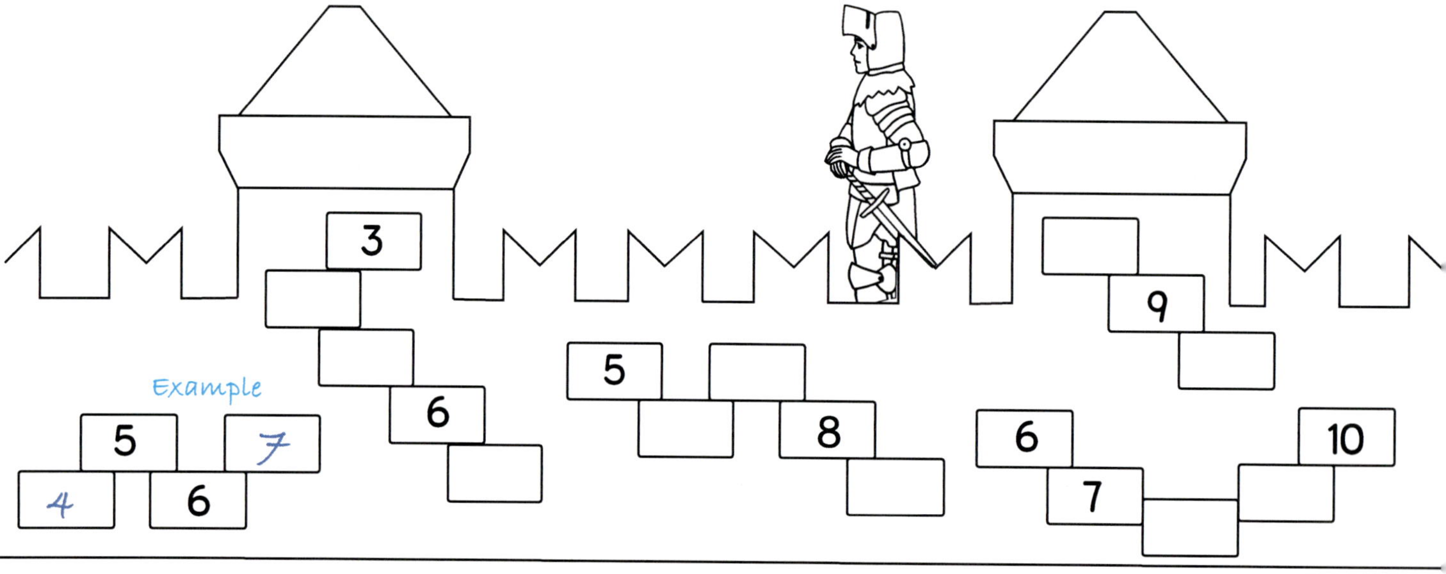

Which tower has more bricks?

put more (>), less (<), or equal (=) signs between the sets of bricks:

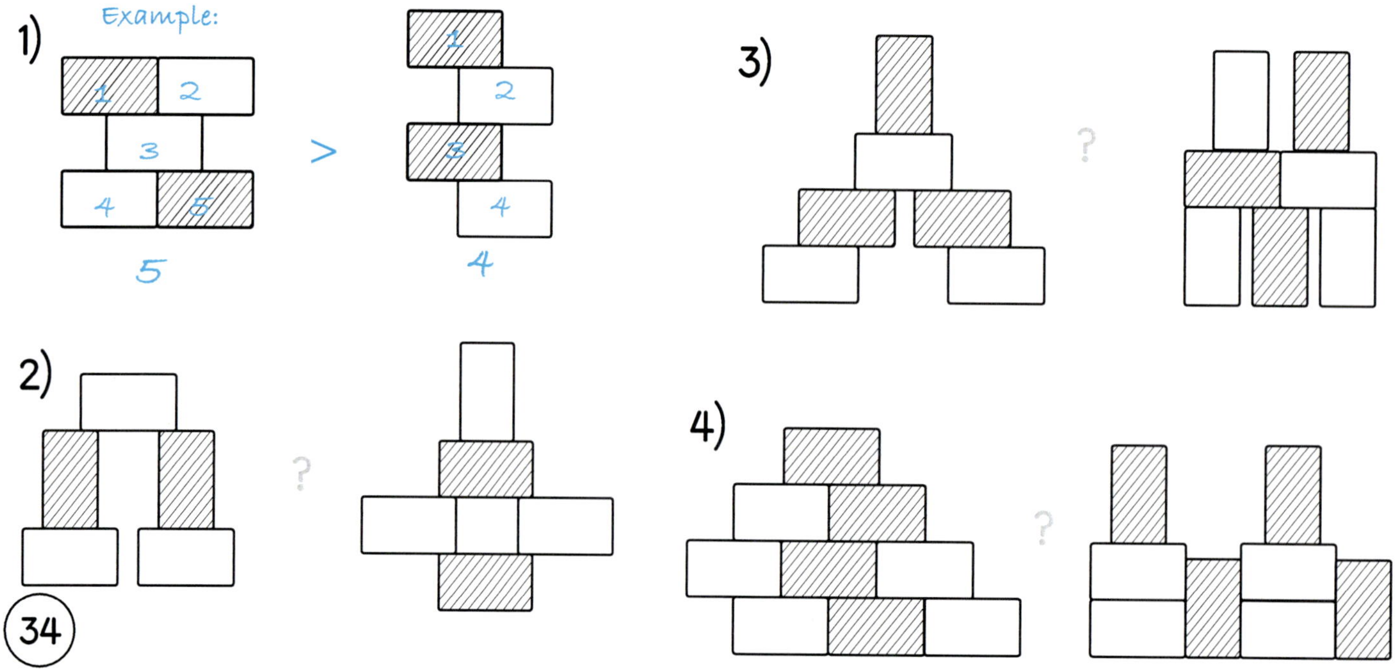

Draw each picture's mirror reflection:

Example

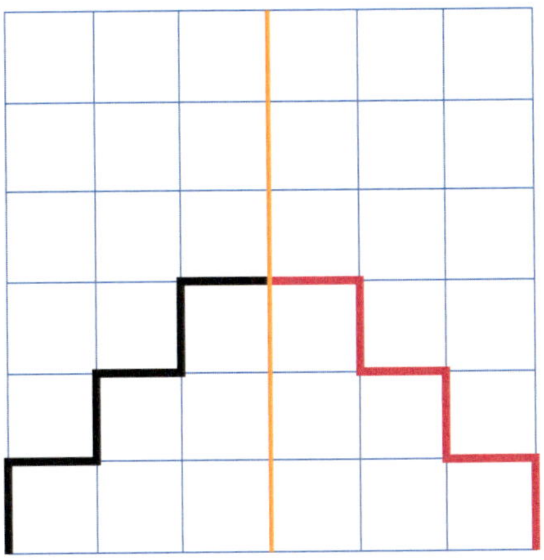

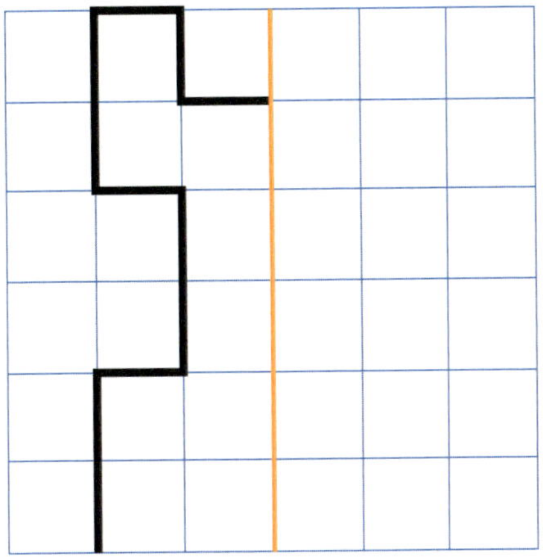

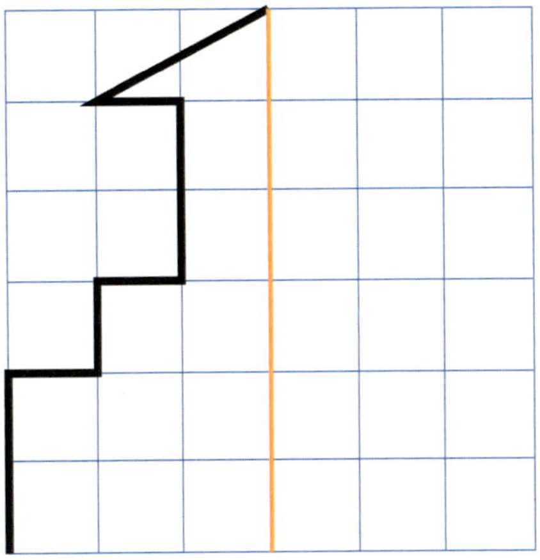

Example

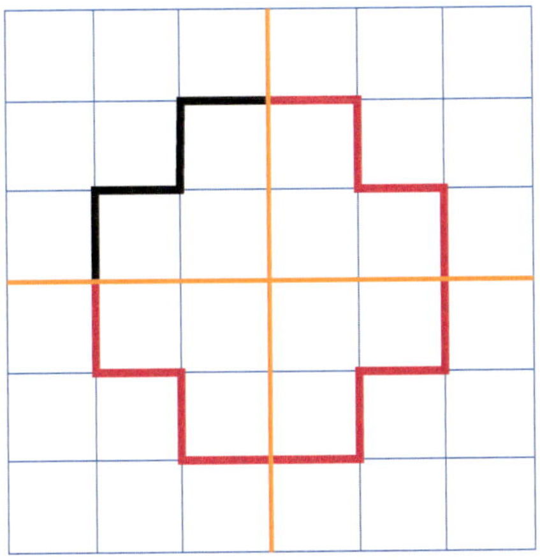

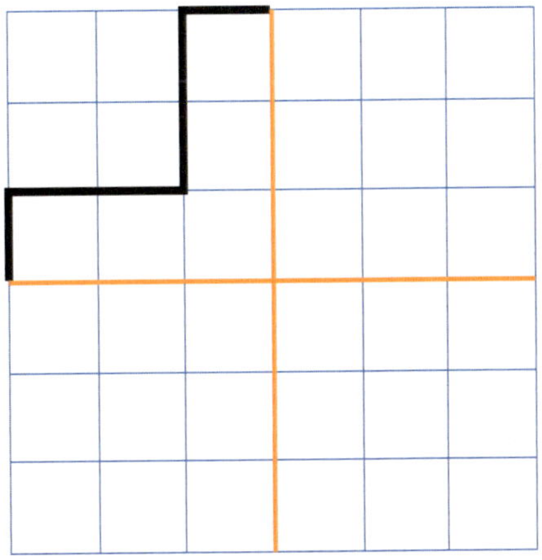

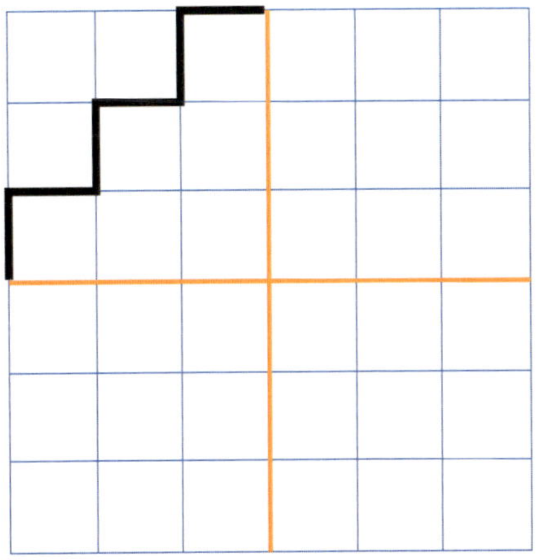

Help the knight find his way to the princess:

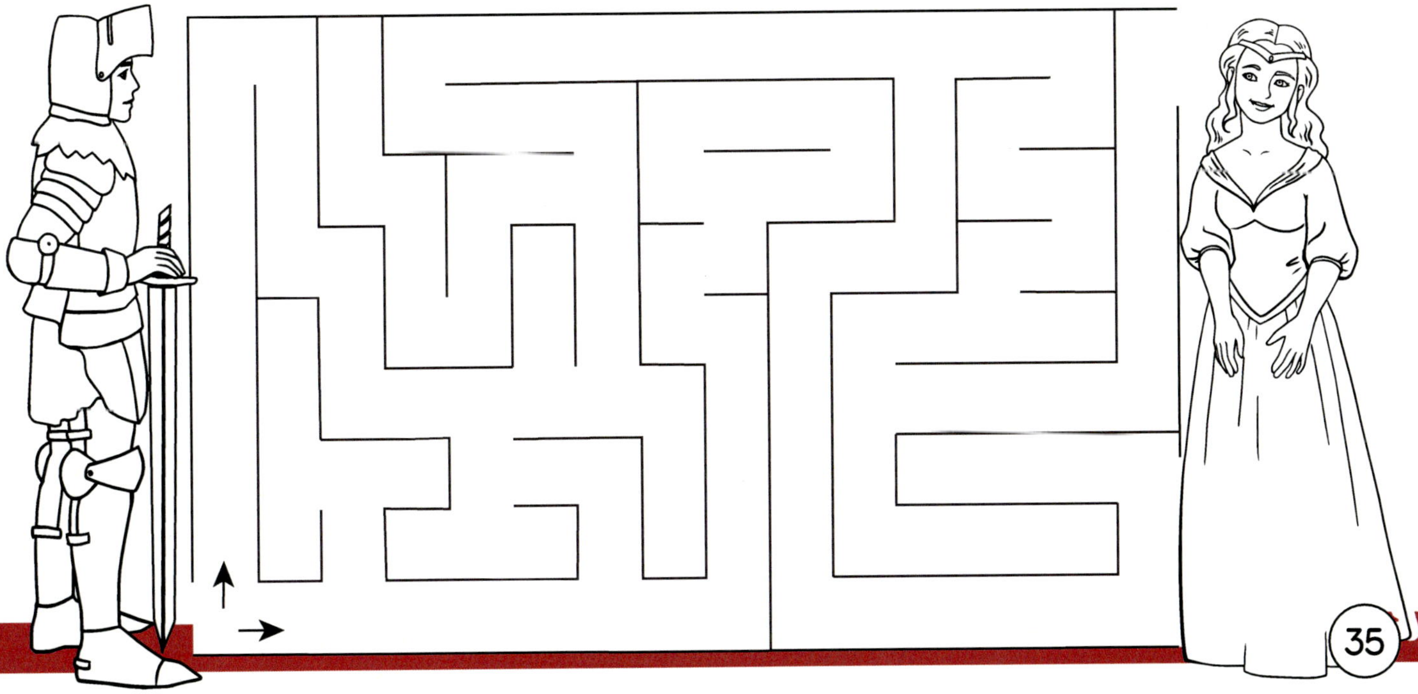

Birthday Party

Date:

Write on the empty flags the missing numbers that are counting backwards:

10, 9, 8, 7, 6, 5, 4, 3, 2, 1, 0.

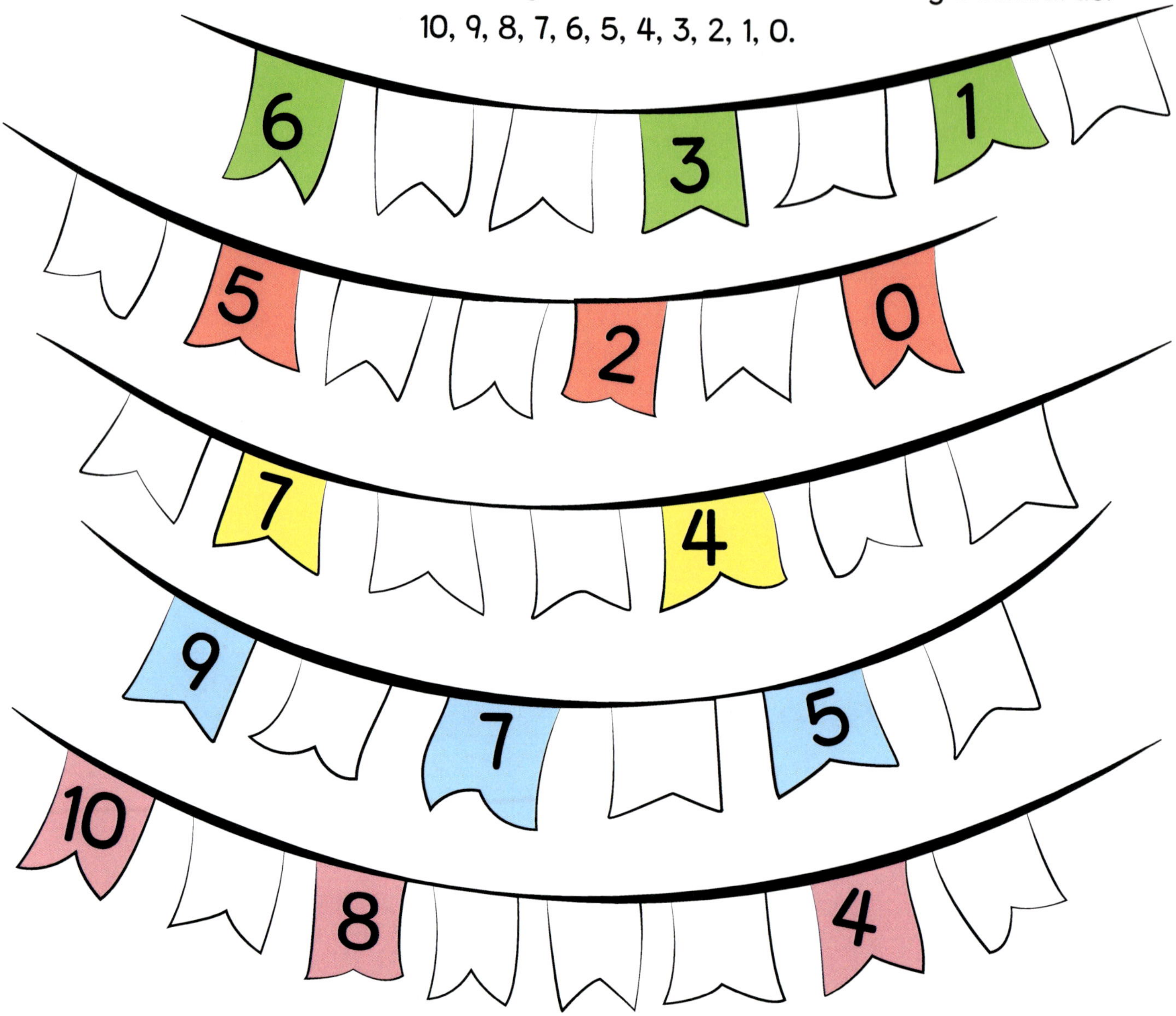

Connect each cake with its missing piece:

Continue the pattern:

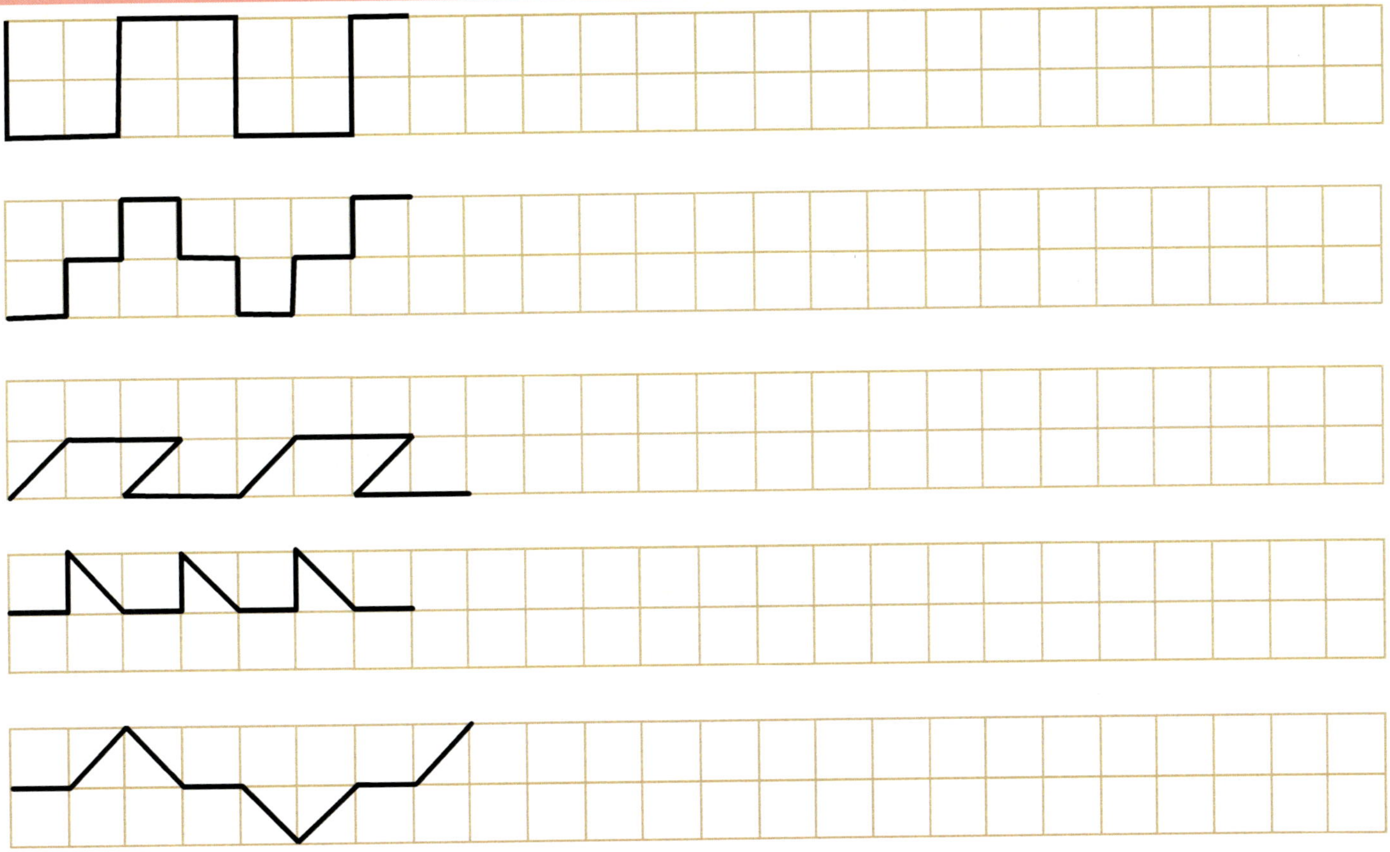

Add more (>), less (<), or equal (=) signs:

0 ? 2	7 ? 0	9 ? 6
3 ? 1	4 ? 8	10 ? 8
2 ? 4	8 ? 8	2 ? 7
5 ? 5	5 ? 2	7 ? 9
6 ? 4	9 ? 8	9 ? 10
3 ? 6	7 ? 7	10 ? 10

Lesson 17: Addition — Fish

Date:

Repeat the numbers below in every aquarium:

1	2	3	4	5	6	7	8	9	10
1									

10	9	8	7	6	5	4	3	2	1

Count the fish.

Example: 5 fish  and  3 fish together is 8

3 fish  and  ? fish together is ?

2 fish  and  ? fish together is ?

3 fish  and  ? fish together is ?

4 fish  and  ? fish together is ?

8 fish  and  ? fish together is ?

Connect all the dots one by one following the numbers.

What did you get?

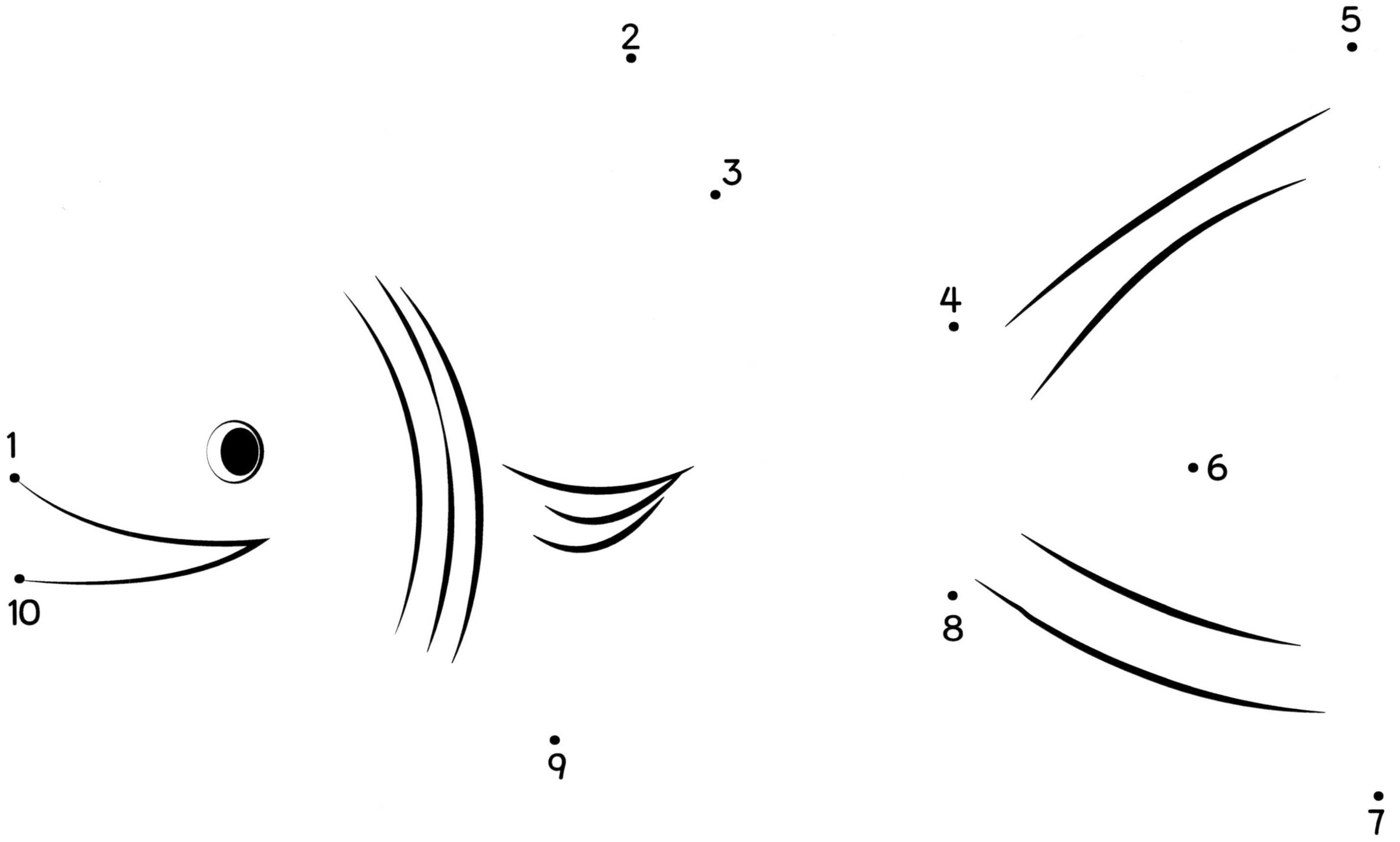

Shadow Match.

Anna got ten orders and needs to deliver ten cakes to her customers, but all the boxes got messed up!

Please help Anna to match each cake to its box.